Innovation – Flood Protection from Heavy Rain, Dam Failures, and Tsunamis

These are not statically calculated construction plans, but rather innovation ideas, the principles of which are presented here.

The contribution is based on existing data and is supplemented by my innovative ideas.

Further open innovations can be found on my page: https://wilkeofficebildungberatung.de/open-innovations-and-more.

The maximum values provided by the AI in January 2025 are also taken into account.

1. Heavy Rain – Maximum Values
China, Zhengzhou (2021):
Precipitation: Over 200 mm in one hour (totaling approximately 600 mm in 24 hours).

Flooding: Parts of the city were submerged up to 3 meters deep.

Cause: An extremely intense rainfall event triggered by a stationary low-pressure system.

India, Mumbai (2005):
Precipitation: 944 mm of rain within 24 hours.

Flooding: Up to 5 meters deep in some neighborhoods.

Cause: Monsoon rains, inadequate drainage, and urbanization.

2. Dam Failures – Extreme Cases
Banqiao Dam, China (1975):
Extent: The Banqiao Dam broke after extreme rainfall (about 1,631 mm over three days).

Flood Wave: The flood reached heights of up to 10 meters and moved at about 50 km/h through surrounding cities.

Damage: Approximately 26,000 people died directly from the flood, with over 200,000 additional deaths due to diseases and hunger.

Johnstown, Pennsylvania, USA (1889):
Extent: The South Fork Dam broke, releasing around 20 million tons of water.
Flood Wave: Up to 18 meters high, hitting the city with devastating destruction.
Damage: Over 2,200 deaths.

Moa Firefighter Catastrophe, Germany (1613):
In medieval times, cities along rivers (e.g., the Elbe) were often flooded up to 7 meters deep after high water events (medieval sources).

3. Road Protection, Assistance, and Rescue Options

Road Protection
Along the roadside, the curbs act like swimmers, rising above the water. As they rise, they pull insulating walls up between the road and the sidewalk, forming an intermediate wall.

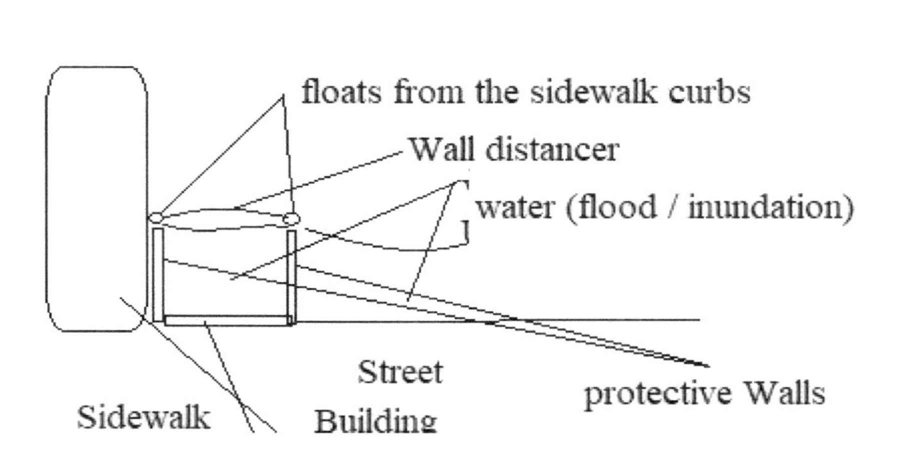

floats from the sidewalk curbs
Wall distancer
water (flood / inundation)
protective Walls
Street
Building
Sidewalk

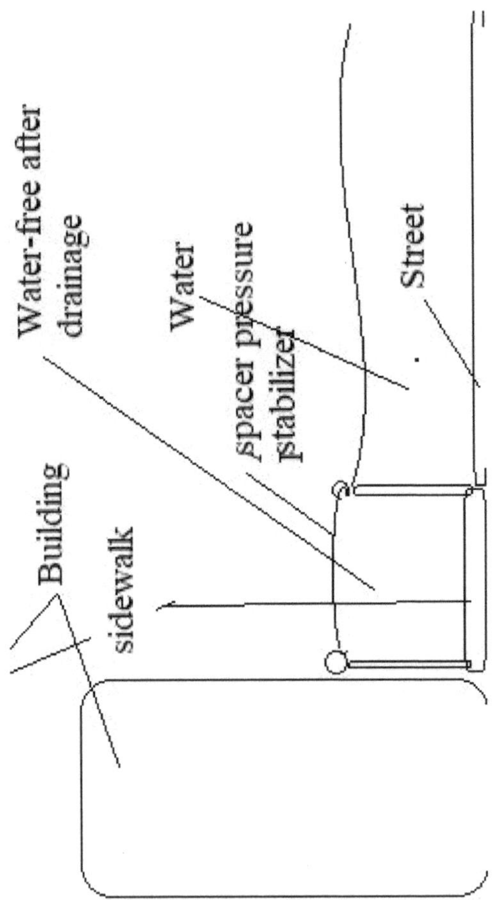

3. Road Protection, Assistance, and Rescue Options

Road Protection

Along the roadside, the curbs act like swimmers, rising above the water. As they rise, they pull insulating walls between the road and the sidewalk upwards, forming an intermediate wall.

At the building wall side, there are rollers between the building walls and the protective wall, acting as pressure buffers.

Both walls are connected at the top like archways to provide stabilization.
Once both walls are raised, water drainage or pumping can begin.

Traffic Islands as Rescue Islands

Traffic islands with a substructure that takes on a floating function. The traffic island rises with the flood level and stays above water. These then function as rescue islands.

To prevent drifting, these are anchored with stainless steel cables.
The steel cables of the smaller islands are optionally located beneath the island.
The road construction is independent of the flood hazard level.
As is well known, street drains allow rainwater to drain off.
To quickly manage such large volumes of water, flap systems are necessary along certain stretches or at specific locations. These flaps open up and direct larger

quantities of water into cisterns.

The flap system is also recommended for train stations that are not underground, or for stations located below ground level.

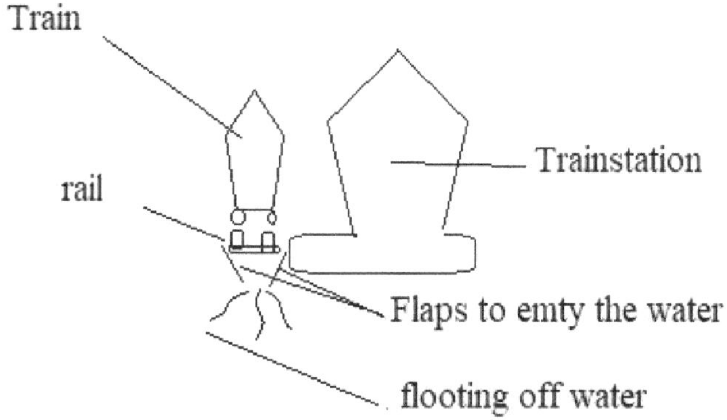

Amphibious Rescue Vehicle (Hovercraft Model)

1. Introduction:
The amphibious rescue vehicle is based on the principle of a hovercraft, which can operate on both land and water. This vehicle is specifically designed for use in disaster situations where rapid and safe evacuations are required, whether in flood areas, over icy surfaces, or in other hard-to-reach locations. Thanks to its hover technology, the vehicle can glide over various surfaces, allowing unrestricted movement in extreme environments.

2. Vehicle Structure:
The vehicle is modular and follows a specific, well-thought-out design that allows to safe transport of those in need of rescue and supplies.

- **Command Area:**
  At the front of the vehicle is a two-person command cabin, which accommodates both the driver and a companion. This cabin is aerodynamically designed to optimize airflow, enabling efficient maneuvering of the vehicle. All necessary controls and communication systems are housed here to ensure safe operation.

- **Supply Area:**
  Directly behind the command zone is a smaller supply area that holds medical packs, drinking water bottles, powdered milk packages, and emergency food packs. This area is specifically designed to provide quick access to aid and supplies that are urgently needed during an evacuation. Rescue teams can quickly access the most essential resources.

- **Basket and Rescue Platform:**
  The rear part of the vehicle is equipped with a flat platform used for evacuating people and animals. On both the left and right sides, there are spaces for up to 4-5 people, allowing evacuees to hold onto the vehicle during transport. The platform is secured with a railing, designed to prevent small children, dogs, or cats from slipping out. The railing is shaped like a basket to enhance safety.

- **Safety Belts and Carabiner Hooks:**
  For additional safety, all individuals on the rescue platform wear a belt with a rope and carabiner hook to secure themselves to the basket railing during transport. This device ensures that no one is at risk, even in strong

currents or rough conditions.
Alternatively, as shown in the image in minibus
form.

3. Functionality and Application:

The vehicle utilizes the hovercraft principle, which
generates an air cushion beneath the vehicle, allowing
for nearly frictionless movement. This enables it to
glide over both solid surfaces and water bodies.
Particularly in flood situations or on difficult-to-access
terrain, the hovercraft model proves invaluable.

- **Quick Evacuation:**
  The vehicle can evacuate people from flood
  zones or remote mountainous areas that cannot
  be reached by conventional vehicles or boats.
  With its ability to travel over both land and
  water, it enables rapid evacuation actions, even
  in extreme weather conditions.

- **Transport of Aid Supplies:**
  The supply area allows the transport of
  emergency relief supplies such as water, food,
  and medication directly to those in need, even if
  they are stranded in hard-to-reach areas or on
  islands amidst floods.

- **Stability and Safety:**
  The hovercraft design ensures stable movement,
  even in turbulent waters or rugged terrain. The
  vehicle's ground clearance allows it to overcome
  obstacles like debris, branches, or small waves
  without hindering its movement.

4. Advantages of the Hovercraft Model:

- **Flexibility:**
  The vehicle can operate in a variety of environments and enables evacuation from areas that are inaccessible to conventional rescue vehicles.

- **Safety:**
  Thanks to the special design of the basket and the safety belts with carabiner hooks, the vehicle is able to transport a large number of evacuees safely, without the risk of injury from slipping or falling out.

- **Ease of Handling:**
  The two-person command cabin allows for simple vehicle control. At the same time, the supply units provide quick and uncomplicated delivery of emergency supplies.

- **Rescue of People and Animals:**
  The design is intended not only to evacuate people but also pets, which is often overlooked in disaster situations.

5. Conclusion:

The amphibious rescue vehicle based on hovercraft technology presents a groundbreaking solution for evacuation and rescue in disaster situations. With its unique combination of land and water capabilities, its safety structure, and its ability to efficiently transport both people and supplies, this vehicle plays a central role in modern disaster relief. It ensures rapid, flexible, and safe evacuations, even under extreme conditions, significantly enhancing rescue capacities during natural disasters.

Expansion of the Amphibious Rescue Vehicle: Airbag Cabin Structure

To further improve the safety and stability of the amphibious rescue vehicle, the design is enhanced with an innovative airbag cabin structure beneath the platform. This structure ensures that even in the event of damage to an airbag, the vehicle remains stable and secure. The air cushion mechanism that carries the vehicle over water and rough terrain is divided into individual, independent cabins. Each of these cabins functions as an autonomous unit, minimizing the risk of total failure in case of a single airbag failure.

1. **Functionality of the Airbag Cabin Structure:**

- **Independent Airbags:**
Beneath the platform of the vehicle, several separate airbags are installed. These airbags are designed to operate independently, meaning that the failure of a single airbag does not affect the overall stability of the vehicle. Each airbag functions like a standalone cabin, bearing the vehicle's weight and keeping it afloat.

- **Safety Mechanism in Case of Damage:**
Should an airbag tear or get damaged—such as by running over a submerged obstacle like a floating debris piece that ruptures an airbag—the load will immediately be taken over by the neighboring cabins. The remaining airbags compensate for the loss, ensuring that the vehicle does not sink or lose stability.

- **Pressure Management:**
Each airbag is equipped with intelligent pressure sensors that monitor the airflow in real-time. If an airbag is damaged, the system can automatically adjust the pressure in the surrounding airbags to maintain the vehicle's stability. This ensures controlled load distribution and prevents the vehicle from tipping or sinking.

2. **Safety Advantages of the Airbag Cabin Structure:**

- **Resilience Against Obstacles:**
When the vehicle drives over a concealed obstacle that damages one or more airbags, the cabin structure ensures that the vehicle does not become immediately unstable. Instead, the remaining cabins take over the role of the

affected airbag, keeping the vehicle stable.

- **Redundancy:**
  If part of the system fails, the overall functionality of the vehicle remains intact due to the redundancy built into the cabin structure. The vehicle can still operate under extreme conditions, which enhances reliability and efficiency during rescue operations.

- **Self-Stabilization:**
  The vehicle automatically stabilizes through dynamic adjustment of the air pressure in the airbags, which is monitored in real-time. This means the rescue mission can continue without delays or sudden instability, even if unexpected damage occurs.

3. Effects on Vehicle Behavior:

- **Improved Maneuverability:**
  With the increased sense of security when crossing obstacles and the independence of the airbags, the vehicle's occupants can act more confidently. The vehicle can also maneuver faster and more precisely, without significant risks to its stability.

- **Land and Water Capability:**
  The cabin structure positively affects the combination of land and water travel by stabilizing the vehicle on both calm water and rough or uneven terrain. This ensures that the vehicle's buoyancy is maintained even in challenging environments.

4. Conclusion of the Expanded Safety Architecture:

The airbag cabin structure represents a significant extension of the safety architecture of the amphibious rescue vehicle, ensuring that the vehicle remains operational even under extreme conditions and in the event of system damage. It ensures that the vehicle is not only resistant to unforeseen damage but also that rescue missions can be carried out safely, even when obstacles or unexpected hazards occur. With the intelligent cabin structure, the vehicle becomes more robust and efficient, significantly expanding its operational capabilities in disaster relief and rescue services.

Big size for out side of citys.

Aerial Rescue Transport Vehicle Comparison

Aerial Rescue Transport Vehicle Comparison

Point	Helicopter	Rotorcraft	Quadcopter
Designation	Classic helicopter, can be versatile	General term for rotor-based flying vehicles	Multirotor flying device, often called a drone
Stability in the Air (at rest, no wind)	Very stable due to large rotors and weight	Very stable due to rotors, but can easily tilt	Stable, but more sensitive with faulty motors
Stability in Strong Wind	Very good, but depends on the model and rotor size	Good to very good, especially in larger machines	Low to medium, highly sensitive to wind
Stability in Storm	Good to medium, rotor can swing in very strong winds	Good to medium, depends on rotor size and weight	Low, mostly uncontrollable in a strong storm

Aerial Rescue Transport Vehicle Comparison

Point	Helicopter	Rotorcraft	Quadcopter
Stability in Storm Rain	Very good, cabins and rotors designed for such conditions	Very good, weather-protected, robust machines	Low, can be heavily affected by rain and moisture
Stability in Sandy Storm	Good, but risk of dust accumulation in the rotors	Good, but rotors are also vulnerable to sand and dust	Low, rotors can block and engines can overheat
Stability in Ash-laden Air (e.g., volcanic eruption)	Moderate, ash can block rotors and engines	Moderate, ash can cause blockages and damage	Very low, ash can completely block the engines
Gravitational Resistance (standard value)	High, requires more energy to fly and climb	High, but variable depending on lift capacity	Low, less resistance due to lighter weight
Air Rescue and Supply Capability with No Landing Space	Very good, can land precisely or hover at low height, even in tight spaces	Very good, similar properties to helicopter, but requires space	Limited, as no landing space is available, only suitable for small supply packages
Flight Altitude and Range	High, some models reach up to 6,000 m, range up to 600 km	High, up to 6,000 m and more, similar range	Low, usually under 500 m, range up to 50 km (depending on model)

Innovation Idea to Enhance Air Stability of Quadcopters:

A windbreaker could represent an innovative solution to improve the stability of quadcopters and other aircraft in windy or stormy conditions. The windbreaker would be deployed from the aircraft's casing during flight to redirect the airflow, providing additional lift or creating a stabilizing barrier.

Considerations and Impact of a Windbreaker System:

- **Reduction of Wind Variability:** The windbreaker could help mitigate the effects of strong winds by reducing the wind load on the rotors, thereby increasing the stability of the aircraft.

- **Increased Efficiency:** By redirecting the wind, the windbreaker could act as additional lift, which would reduce the energy consumption of the quadcopter.

- **Improved Maneuverability:** In windy or stormy conditions, the aircraft would be easier to control due to fewer fluctuations caused by the windbreaker system.

- **Expanded Operational Range:** With such a system, quadcopters could be more effectively deployed in adverse weather conditions, particularly in stormy areas or in challenging rescue missions.

The concept of a windbreaker could be particularly beneficial for deploying quadcopters in extreme weather conditions, significantly enhancing their use in air rescue and emergency services.

1. Wind Flow Redirection and Lift Enhancement

- **Principle of Lift Flow Redirection:**
  A windbreaker that unfolds to redirect the wind flow around the gondola or basket of the aircraft could theoretically help minimize the wind's impact on the aircraft. By redirecting the airflow, the wind does not directly hit the gondola or basket, reducing the oscillations or vibrations of the aircraft. This technique would be especially beneficial in turbulent or windy conditions, as it reduces the direct influence of wind on the structure of the aircraft, improving stability and control.

- **Additional Lift:**
  If the windbreaker works similarly to a **wind wing**, it could generate additional **lift** by directing the airflow in a favorable direction for the aircraft. This could be particularly useful when the aircraft is facing **headwinds** or **crosswinds**. The windbreaker would help guide the wind around the sensitive parts of the aircraft (like the gondola or basket), thus reducing the **sideways swing** that could destabilize the aircraft. The wind would flow over the aircraft more smoothly, reducing the vibrations caused by wind pressure on the gondola.

- **Effective Use of Wind:**
  Particularly in **headwinds** or **crosswinds**, the windbreaker system would help redirect the airflow in a **beneficial direction**. This would prevent the wind from directly impacting the

sensitive areas of the aircraft, such as the gondola, basket, or engine covers. Instead, the airflow would pass over the aircraft, avoiding direct contact with the structure and thus reducing **oscillations** or **vibrations**. This technique would not only improve the airflow around the aircraft but also contribute to optimizing the **aerodynamics** of the entire system, leading to better control and more efficient flight.

In summary, **wind flow redirection** through a windbreaker would not only stabilize lift and reduce oscillations but also improve the aerodynamic properties by directing the wind smoothly over the aircraft. The result would be enhanced **flight stability** and **better flight performance** in challenging wind conditions.

2. Reduction of Fluctuations and Turbulence

- **Fluctuation Reduction:** The main advantage of a windbreaker system would be the reduction of fluctuations caused by sudden air currents or turbulence. Especially in situations where the wind shifts abruptly or changes rapidly, the deployed windbreakers could help keep the aircraft more stable. This would minimize the influence of crosswinds or sudden gusts, thereby improving control and safety.

- **Turbulence Management:** A windbreaker could serve as a buffer zone in turbulent conditions, distributing a larger airflow and directing it more smoothly over the gondola or the basket. This would reduce the sudden air vortices or oscillations that could throw the aircraft off balance.

The windbreaker would thus help to manage and mitigate the challenges posed by dynamic air conditions, making the aircraft easier to control and more stable in turbulent or unpredictable weather.

3. **Aerodynamic load and additional complexity**

- **Additional aerodynamic load:** Such a windbreaker would affect the aerodynamic efficiency of the aircraft. While on the one hand it can help increase stability, on the other hand it could affect drag and performance as more components and structures have to work in the wind. Especially at higher speeds or in stormy conditions, this additional drag could reduce the performance of the aircraft.

- **Mechanical load:** A windbreaker would also create additional mechanical load, especially when it is deployed and retracted. The system would have to be designed to be both robust enough to function in extreme wind conditions and light enough not to excessively increase the overall mass and weight of the aircraft.

- **Complexity and maintenance:** Introducing a windbreaker system would introduce additional complexity to the design and maintenance of the aircraft. The mechanism would have to be able to be deployed reliably and quickly without affecting the functioning of the rotors or other flight systems. The possibility of malfunctions or defects in mechanical parts must also be taken into account, which would require regular maintenance and testing.

4. Integration into Existing Aircraft

- **Helicopters/Helicopter-like Aircraft:** For larger machines, a windbreaker that enhances stability during strong winds or storms could be an interesting addition. Especially in rescue operations or when transporting in difficult weather conditions, this could help keep the gondola calmer and make control easier. However, the windbreaker must be designed to avoid negatively affecting the rotors or the overall stability of the aircraft.

- **Quadcopters:** For smaller multirotor devices like quadcopters, a windbreaker could also be beneficial, particularly if it reduces fluctuations caused by crosswinds or turbulence. However, the size and weight of the system would be crucial, as quadcopters have limited payload capacity. A windbreaker that is too heavy or complex could negatively affect flight performance.

Conclusion

A windbreaker that deploys as a stabilizing measure in aircraft and redirects the airflow could be a promising solution, especially in extreme weather conditions such as strong winds, storms, or turbulence. However, the technology would come with challenges, such as balancing aerodynamic efficiency, mechanical stress, and the added complexity of the design.

The design would need to be optimized to ensure that the windbreaker works effectively without compromising the overall performance of the aircraft. For larger machines like helicopters and larger drones, this could be a practical addition, while for smaller quadcopters, implementing such a system could be problematic, especially if it restricts weight or maneuverability too much.

Windbreaker for Quadcopters Using a Combination of Materials (Chitin and Aluminum Skeleton):

- **Material Choice:** Using a combination of synthetic chitin and an aluminum skeleton for the windbreaker could indeed provide a good balance between stability and weight. Chitin, being lightweight yet strong, could form the core of the windbreaker, providing structural integrity, while the aluminum skeleton would give it rigidity and strength, enabling it to withstand wind forces without excessive weight.

- **Benefits of This Material Combination:**

  - **Weight Efficiency:** Chitin is a lightweight material, so it would help minimize the added weight to the quadcopter.
  - **Strength and Durability:** Aluminum, with its ability to resist corrosion and handle mechanical stress, would ensure that the windbreaker remains durable under harsh conditions.
  - **Aerodynamic Flexibility:** The material combination would allow for efficient deployment and retraction, crucial for minimizing drag when the windbreaker is not in use.

This combination could offer a practical and effective solution, as it balances the necessary weight constraints of a quadcopter with the durability and aerodynamic performance needed for stability in turbulent air.

Advantages of the Material Selection for the Windbreaker:

1. **Artificial Chitin:**

    - **Light and Flexible:** Artificial chitin is a very lightweight and flexible material. These properties allow the windbreaker to be designed in such a way that it can be deployed when needed without significantly increasing the overall weight of the quadcopter. This is especially important because quadcopters have limited carrying capacity.
    - **Strength:** Artificial chitin offers high strength relative to its weight. This ensures that the windbreaker remains stable and maintains its shape without adding unnecessary load to the aircraft.
    - **Resistant to Weather Influences:** Artificial chitin is resistant to moisture and UV radiation, making it reliable in fluctuating weather conditions such as rain, sun, or strong winds. It is also relatively abrasion-resistant, which could extend the lifespan of the system, especially with frequent use under varying environmental conditions.

2. **Aluminum Skeleton:**

    - **Robust and Light:** Aluminum is an ideal material for the frame of the windbreaker as it is both robust and lightweight. This combination allows for the creation of a strong, stable skeleton without unnecessarily weighing down the aircraft. This is especially important for a quadcopter, where payload capacity is critical to flight performance.

- **High Strength:** Aluminum provides excellent strength and can withstand wind pressure, mechanical forces during deployment, and the wind speeds experienced without bending or breaking. This ensures that the skeleton of the windbreaker maintains its structural integrity even under challenging conditions.
- **Good Workability:** Aluminum is easy to process, allowing for precise design of the windbreaker system. This flexibility in construction is important for creating a system that fits seamlessly into the existing design of the quadcopter, especially for a deployable or retractable system. The material is therefore excellent for producing mechanical components that need to operate reliably and functionally.

Conclusion: The combination of artificial chitin and aluminum offers an excellent balance between lightness, flexibility, stability, and durability. The windbreaker would be able to effectively redirect airflow around the aircraft while keeping weight low and maintaining resilience against external influences. This material pair is particularly suitable for achieving the desired benefits of a windbreaker without negatively affecting the overall performance of the quadcopter.

Design and Functionality of the Windbreaker System:

1. **Deployment Mechanism:**

   - **Aluminum Skeleton as a Stable Framework:** The aluminum skeleton would provide a solid structure to support the artificial chitin elements when deployed. The windbreaker could be activated via a spring mechanism or a small electric motor, allowing the segments to be quickly and efficiently extended when needed.

   - **Artificial Chitin Elements as Folding Segments:** The artificial chitin elements would act as panels or segments that unfold along the aluminum skeleton. When not in use, these elements would be stored compactly within the quadcopter, ensuring that no additional aerodynamic drag is created when the windbreaker is not deployed.

2. **Aerodynamics and Function in Wind:**

   - **Wind Redirection:** When deployed, the windbreaker would be designed to redirect the airflow in such a way that it shields the gondola or basket of the quadcopter. The shape of the windbreaker should be optimized to channel the airflow smoothly over the aircraft, avoiding turbulent eddies that could negatively affect the stability.

- **Prevention of Wind Fluctuations:** The use of the windbreaker structure would help keep the flight steadier, especially in strong winds or sudden gusts. The goal is not only to stabilize lift through the rotors but also to stabilize the airflow around the quadcopter, thereby reducing sway and improving control. The redirection of wind would contribute to reducing the side-to-side movement of the vehicle, helping it stay more stable during turbulent conditions.

This system would offer a practical solution to improving the stability and handling of quadcopters in challenging weather conditions, without significantly increasing the weight or affecting overall performance. The combination of a deployable, lightweight windbreaker and the aerodynamic benefits of wind redirection could make flying in turbulent or gusty winds safer and more efficient.

3. Folding Mechanism:

- **Compact and Space-Saving:** Given that the quadcopter has limited space for additional equipment, the windbreaker must fold up as compactly as possible when not in use. A clever folding mechanism could ensure that the windbreaker panels are stored in a zigzag pattern or as a parabolic surface within a small housing underneath the aircraft. This would minimize the footprint and avoid excess drag when the system is retracted.

- **Simple Deployment and Retraction:** The mechanism for deploying and retracting the windbreaker should be straightforward and reliable. A small motor could automatically activate the system when wind speed exceeds a set threshold. Alternatively, the windbreaker could be manually deployed by the pilot if they deem it necessary, providing flexibility in operation.

4. Durability Against External Influences:

- **Turbulent Air and Rain:** Artificial chitin is not only strong but also relatively resistant to moisture. This means that the material would retain its strength even in rainy or wet conditions, ensuring that the windbreaker remains effective even in adverse weather.

- **Protection Against Mechanical Damage:** The aluminum skeleton provides additional security by acting as a shock absorber. This feature could prevent the windbreaker from being damaged during intense turbulence or potential collisions, making it more robust and reliable in extreme conditions.

This combination of a compact, easy-to-deploy windbreaker with robust materials would ensure that the system remains effective and durable, even in challenging flight environments. The balance between aerodynamics, durability, and space efficiency would make it a viable addition to the quadcopter's design, enhancing stability and control in turbulent weather.

Performance and Challenges:

1. Performance Optimization:

- **Wind Redirection Efficiency:** The windbreaker's effectiveness would depend on how well it can redirect the airflow without creating additional drag or burdening the rotors. An overly heavy or poorly designed system could lead to inefficiency, reducing the quadcopter's flight time and performance. By using materials like artificial chitin and aluminum, the system can remain lightweight while being functional and stable, optimizing flight performance.

- **System Weight:** The weight of the system plays a crucial role in maintaining the quadcopter's overall efficiency. With the selected materials, the windbreaker can be kept light enough to not interfere with the quadcopter's lift and propulsion systems.

2. Challenges:

- **Complexity of Construction:** The mechanism needs to be reliable and easy to operate to prevent issues during deployment and retraction. For smaller quadcopters, space limitations could become a significant factor, making it necessary to create a compact and effective folding mechanism.

- **Cost and Manufacturing:** Producing such a windbreaker system could be expensive, as it requires high precision in manufacturing and specialized materials. The combination of lightweight and durable materials like artificial chitin and aluminum may add to the cost of the system.

- **Wind Strengths:** The windbreaker may lose some of its effectiveness in extreme weather conditions, such as very high wind gusts or turbulent airflow. While it provides added stabilization, it doesn't guarantee basic flight safety, which is dependent on the quadcopter's overall design and capabilities.

Conclusion:

A windbreaker made of artificial chitin and an aluminum skeleton is an innovative approach to improving stability in windy conditions for quadcopters. It can effectively channel airflow around the quadcopter, reducing vibrations and enhancing stability in stormy or turbulent conditions. The combination of artificial chitin and aluminum provides a good balance of lightweight construction, strength, and resilience against external factors.

Additionally, incorporating a **telescope frame structure with an inflatable mechanism** controlled by wind and magnetic field sensors could provide dynamic adjustments to the system. This concept would combine both aerodynamic and magnetic technologies to dynamically control the frame's shape, allowing the system to adapt to various wind conditions in real-time, enhancing stability even further.

Let's explore this concept in detail:

Inflatable Mechanism and Magnetic Field Sensor Integration:

1. **Telescope Frame Structure:**

   - The windbreaker system could feature telescopic segments that extend outward when needed. These segments would be lightweight, ensuring minimal impact on the quadcopter's flight time and maneuverability. The telescopic design would allow the system to collapse neatly when not in use.

2. **Inflatable Mechanism:**

   - The inflatable components would allow for quick expansion of the windbreaker's surface area when necessary, creating a more significant barrier against wind. The airbladders would be made from a durable, lightweight material that inflates quickly when activated by wind sensors or manually by the pilot.

   - The inflation could be triggered automatically by sensors that detect wind speed, direction, and intensity, or could be manually controlled by the pilot, depending on the system's configuration.

3. **Magnetic Field Sensors:**

   - The integration of magnetic field sensors could help dynamically adjust the shape of the windbreaker frame. These sensors could detect variations in the magnetic fields caused by wind flow and adjust the tension or shape of the inflatable components to optimize wind redirection.

4. **Dynamic Control and Aerodynamics:**

- The combination of inflatable elements and magnetic field sensors would allow the system to adapt dynamically to changes in the weather. When the wind shifts direction or increases in intensity, the frame would adjust to redirect the airflow smoothly around the quadcopter, maintaining stability and reducing turbulence.

This dynamic system would not only offer improved performance but also provide a more adaptable and responsive windbreaker solution, allowing for greater flexibility and control in challenging conditions. The ability to adjust the windbreaker's shape and size in real-time ensures optimal performance across a wider range of wind speeds and directions.

Description of the Concept Sketch for the Telescopic Windbreak for a Quadcopter

1. **General Shape of the Quadcopter:**

- **Quadcopter Gondola:** At the center of the sketch is the quadcopter with four rotors (two at the front, two at the rear). The gondola is the central area that houses the rotors, control electronics, and payload (e.g., rescue equipment).

-

2. **Telescopic Windbreak Frame:**

- **Frame Structure:** The windbreak is designed as a telescopic system that can be extended when needed. To utilize the wind direction and stability, the frame surrounds the gondola of the quadcopter. The frame consists of several rings stacked on top of each other, which interlock when extended.
- The outer side of each frame is made of lightweight, flexible material (such as Kunstchitin), while the inner skeleton is made of aluminum for additional stability. The telescopic rings are equipped with locking mechanisms that automatically snap into place when extended.

- **Air Chambers in Each Ring:** These chambers fill with air when the wind blows, stabilizing the windbreak and helping to control the lift and airflow around the quadcopter.

3. **Inflation Mechanism:**

- **Air Hoses or Valves:** Air valves or hoses are depicted at the edges of each telescopic ring, allowing wind to flow through these channels and fill the chambers. When the wind blows, it enters through these passages and inflates the chambers, stabilizing the structure.
- A pressure sensor is located at the center

of the structure, measuring the air pressure and ensuring that the chambers are not overfilled. The regulation mechanisms help manage wind pressure.

4. **Magnetic Field Sensors and Shape Control:**

   - Magnetic field sensors are integrated around the telescopic frame in small boxes or modules. These sensors monitor the magnetic field and send necessary impulses to the AI, which then adjusts the shape of the frame.
   - **Shape Adjustment:** The sensors detect wind direction and magnetic field strength and send corresponding impulses to the metal reinforcements in the inner skeleton. This allows the frame to dynamically adjust, ensuring it always takes the optimal shape for the current wind conditions.

5. **Unfolded Structure:**

   - The structure of the windbreak is shown in the "open" position. The telescopic rings are fully extended and filled with air. The chambers are visibly inflated, giving the frame a stable, aerodynamic shape that redirects the wind and stabilizes the quadcopter.

6. **Position of Sensors and Control Unit:**

   - In the sketch, the control unit for the windbreak as well as the sensors could be clearly depicted as small boxes placed both on the quadcopter itself and within the telescopic structure. These units communicate with the AI, which monitors and adjusts the structure in real time.

7. **Wind Direction and Magnetic Field Influence:**

    - A wind arrow or airflow lines may be drawn on the sketch, indicating the wind direction and strength. The wind stream is redirected by the telescopic rings, which can be shown with lines or arrows along the frame and the gondola.

Summary of the Sketch:

- **Quadcopter with Telescopic Windbreak:** At the center of the image is the quadcopter with a square or rectangular frame. Surrounding it are the extended telescopic rings.
- **Telescopic Frame:** The rings are stacked on top of each other, reinforced with artifacle chitin on the outside and aluminum on the inside.
- **Air Chambers:** The edges of the rings are equipped with chambers that fill with air when the wind blows to stabilize the frame.
- **Magnetic Field Sensors and AI Control:** Small boxes or modules around the frame and gondola, equipped with sensors that monitor the magnetic field and airflow.
- **Shape Adjustment and Stabilization:** The windbreak has a dynamic shape that changes based on wind strength and direction.

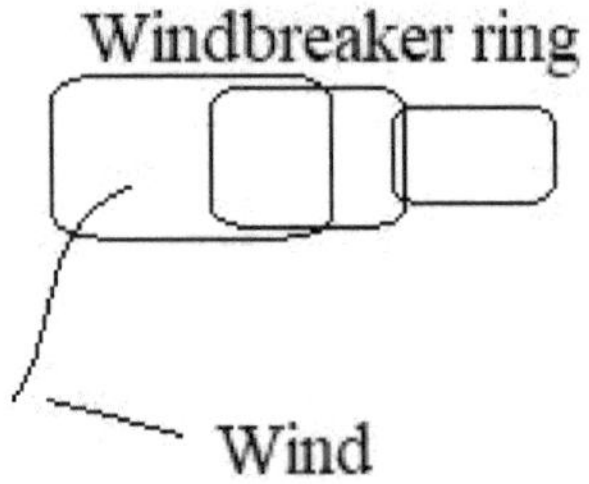

Teleskop arms from the Quadrocopter to the Windbreaker ring
Wind

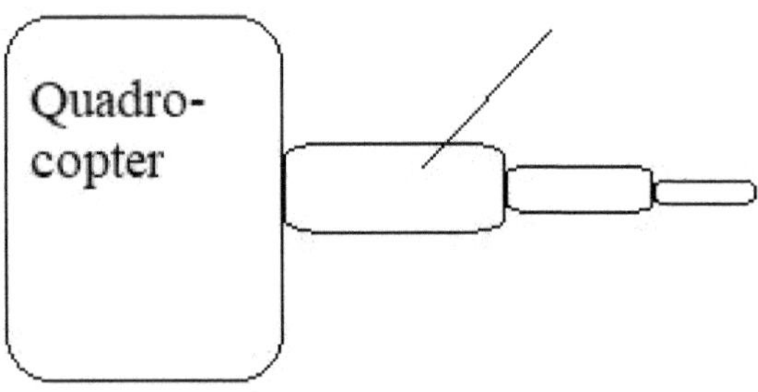

Wind filled telescope arm
Quadro-copter

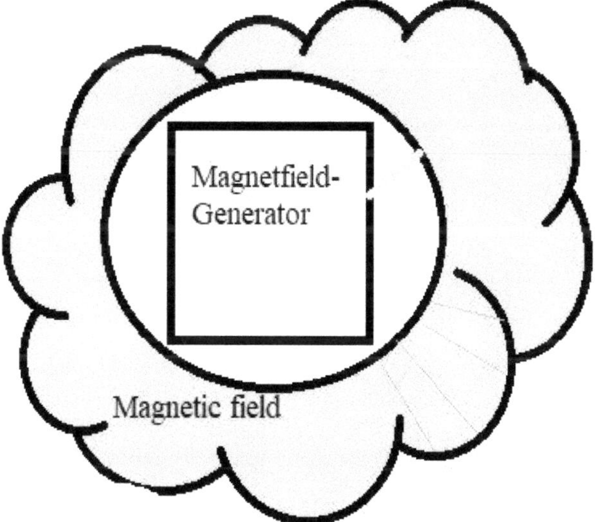

Shaping by different strengths of magnetic field impulses

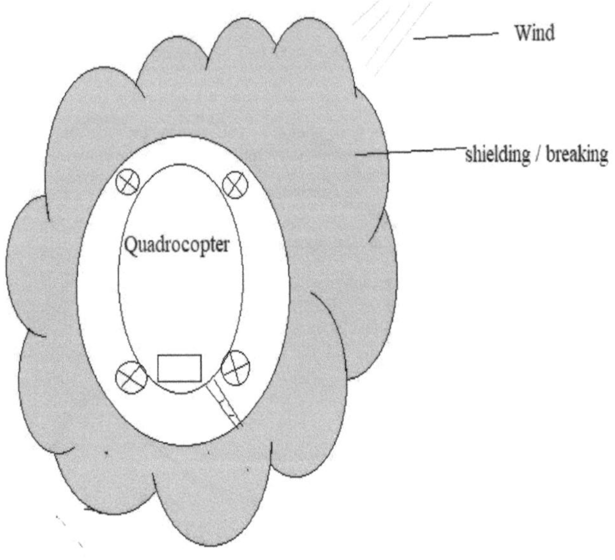

The principle image does not show correct aerodynamics.

The interlocking mechanism is envisioned like the shield wall of the Roman legions, where the shields (scales) interlock with each other. The scale attachment is movable, allowing the scales to form a shape similar to a water surface, which is controlled by a regulated magnetic field.

Dam with Early Warning System

Many dams are located near cities around the world. The two most significant known risks are flooding or dam failures.
Most dams are already equipped with early warning systems.
Dams are typically built to keep areas dry or for energy and drinking water generation.
If the dams are built to keep an area dry, it usually

involves redirecting a river from its natural course. The second type involves a moderated continuation of the river with a reservoir directly after the dam.

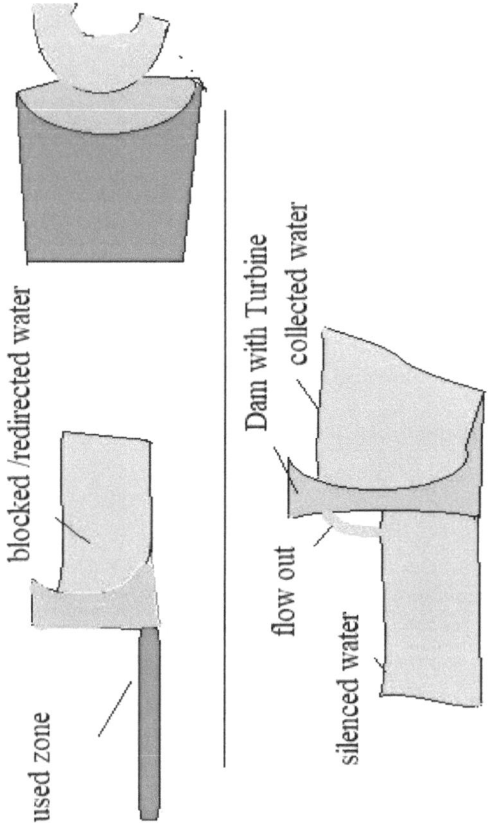

As the water level rises above the normal level, turbines are installed in four stages. (The turbine generates energy through the water flow of the redirected water. In this case, the turbines are installed horizontally on the dam wall at three different safety heights. The turbine generates energy for the signal

horn. The signal tone changes as more turbine levels become active. This informs the residents about the current level reached. The fourth turbine signal indicates that the water is flowing over the dam.)
The signal is accompanied by radio announcements and traffic information boards.

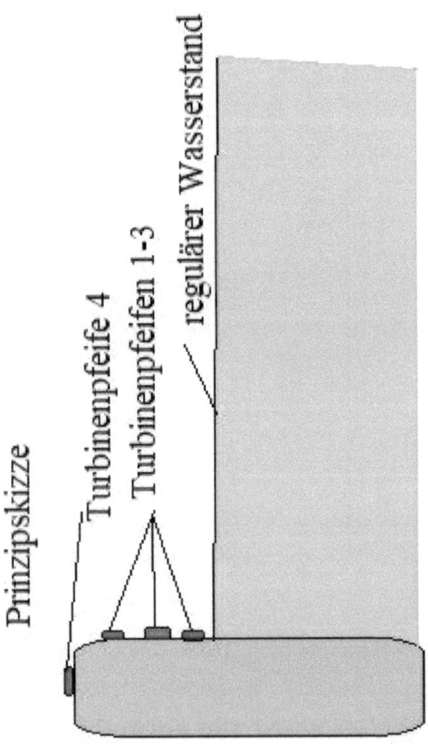

A second warning variant is the activation of float switches.

If the water level rises above the reference value, the water lifts the float, which then activates the signal horn. The floats are installed in the same way as the turbine pipes.

The fourth float also indicates that the water is flowing over the dam. Therefore, it should be placed in a kind of depression in the dam so that the float rises even with a small water overflow.

tsunami protection variant

early warning system

last strongest 4 tsunamis

Date	Location	Water retreat (Meter)	wave height (Meter)	wave force (kN/m²)
26.12.2004	Indian Ocean (Sumatra)	up to 1,5	up to 30	up to 30
11.03.2011	Tōhoku, Japan	up to 1,2	up to 40	up to 40
28.09.2018	Sulawesi, Indonesia	up to 1,0	up to 7	up to 7
16.09.2023	Dickson Fjord, Greenland	unknown	more then 200	unknown

The massive amount of water for the tsunami wave barrier originates from the displacement of water due to sudden movements of the seabed. This displacement primarily occurs through:

1. **Seismic Activity (Earthquakes)**
   When an earthquake raises or lowers the seabed, the water above it is vertically displaced. Gravity compensates for this displacement by causing the water to flow radially away from the earthquake's epicenter, generating waves in the process.

2. **Underwater Landslides**
   Landslides or volcanic activities under water also displace water, which then becomes visible as tsunami waves.

3. **Displacement of Large Amounts of Water**
   The entire water column above an earthquake source moves. This explains why even distant coastlines are affected by the wave energy.

The seismic activity from December 16 to 26, 2004, was particularly significant, as it included the devastating earthquake of December 26, 2004, in the Indian Ocean, which triggered a massive tsunami.

Tōhoku Earthquake and Tsunami – March 11, 2011

- **Date and Time**: March 11, 2011, at 14:46 JST (05:46 UTC).
- **Location**: Off the east coast of Tōhoku, Japan.
- **Magnitude**: 9.1 (one of the strongest earthquakes in recorded history).
- **Epicenter**:
  - Approximately 72 kilometers (about 39 miles) east of the Oshika Peninsula,

Miyagi Prefecture, Japan.
  - Coordinates: 38.297° N, 142.372° E.
- **Depth**: 29 kilometers below the ocean floor.
- **Tsunami Impact**:
  - **Wave Heights**: Up to 40 meters in certain coastal areas.
  - **Affected Areas**: Mainly the northeastern coast of Japan. The tsunami flooded several prefectures, including Miyagi, Iwate, and Fukushima.
  - **Consequences**: The tsunami waves eventually reached other Pacific Rim countries. The devastation in Fukushima led to a nuclear meltdown in several reactors at the Fukushima Daiichi Nuclear Power Plant.

Connection with Seismic Activity:
The Tōhoku earthquake was caused by the subduction of the Pacific Plate beneath the North American Plate. The movement along a 500-kilometer-long fault zone led to a massive uplift of the ocean floor and the resulting tsunami.

Tsunami Early Warning Systems:

1. **Oceanic Tsunami Early Warning Systems:**

   - **Seismic Stations**: Since most tsunamis are triggered by underwater earthquakes, early warning systems rely on detecting seismic tremors. A global network of seismic stations is installed around the Pacific and other tsunami-prone oceans.

- **Deep-water Buoys**: These buoys measure changes in water depth and can detect tsunami waves before they reach coastal areas. They send immediate data to tsunami warning centers.
- **Satellites**: New technologies also use satellites to monitor tsunamis in real-time, which improves the precision and reach of early warnings.

2. **Coastal Warning Systems:**

- Many tsunami-prone regions have local warning systems to enable timely evacuation. These systems broadcast alerts through loudspeakers, SMS, radio, and TV.
- **Apps and Mobile Alerts**: In many countries, apps are now available that are directly linked to tsunami early warning systems, sending immediate local warnings to residents' smartphones.

3. **International Collaboration:**

- The **Intergovernmental Oceanographic Commission (IOC)** of UNESCO has set up a coordinated tsunami early warning platform worldwide, facilitating the rapid exchange of information between affected countries.
- Particularly in the **Indian Ocean** and **Pacific**, well-established systems are regularly tested to ensure early warnings function even during smaller

earthquakes.

Damage Mitigation Measures:

1. **Timely Evacuation:**

   - The primary measure for mitigating tsunami damage is the **evacuation** of coastal areas before the tsunami waves arrive. To this end, there are clear **evacuation plans** and well-marked **tsunami evacuation routes**.
   - **Tsunami Evacuation Centers**: Many coastal cities have higher buildings or designated emergency centers that serve as safe havens.

2. **Infrastructure and Coastal Protection:**

   - **Tsunami Barriers**: In countries like Japan and Indonesia, there are massive protective walls and dams designed to reduce the impact of smaller tsunamis. However, they cannot protect against all tsunamis, especially those with extremely high waves.

   - **Tsunami-Resilient Building Design**: In many coastal regions, new buildings are now being constructed with tsunami resilience in mind, featuring higher foundations and durable materials.

3. **Emergency Preparedness:**

   - **Emergency Kits**: People living in

vulnerable areas are encouraged to prepare emergency kits with food, water, first aid supplies, and other essential items.
- **Training and Drills**: Regular **tsunami drills** are conducted in many vulnerable areas to ensure the population is prepared for a real tsunami event.

4. **After the Tsunami:**

   - **Rapid Response and Rescue Operations**: After a tsunami, emergency services must act quickly to rescue survivors and provide medical assistance. Countries with tsunami risks have specialized teams trained for such disasters.
   - **Rebuilding**: The rebuilding process after a tsunami is often lengthy and expensive. Countries like Japan and Chile have extensive programs to support affected regions, both financially and in rebuilding infrastructure.

Challenges:

- **Early Warning Times**: In some regions, such as the Caribbean or the Indian Ocean, tsunamis are often difficult to predict, as they can also be triggered by landslides or volcanic activity. In such cases, the warning time is very short, requiring quick responses.
- **Climate Change**: Rising sea levels could amplify the effects of tsunamis in the future, as coastal protection measures become less effective. Despite these challenges, research has

made significant progress in recent years, and the global early warning system has improved considerably. However, the greatest challenge still remains to assist the population quickly and effectively.

Current Measures: Tsunami barriers are already being used, but they are quite costly to build and maintain. These barriers are designed to weaken the power of tsunami waves and, if possible, protect the land they are placed around.

Additional Idea!!! by me

Stage 1: Moat as a Blocker The moat would serve as the first line of defense, dampening the wave's energy and slowing or preventing the wave's advance. The moat could be a deep depression in the ground. This barrier would affect the wave enough to prevent it from crossing the wall, or at least slow it down enough to make the "crossing" more difficult.

Stage 2: Implosive Opening – When – the Wall is Overcome If the wave can overcome the moat but not the wall, the wall could open in an implosive manner. This mechanism would "suck" the wave's pressure and draw the water flow into the wall. Once the wave has completely surpassed the wall, the wall could open (perhaps through a mechanical valve or flap structure), allowing water to flow into reservoirs or chambers within the wall. The water from the wave would then be drawn through the wall and remain inside this structure. The wall would stay intact and be stabilized by the added weight of the water.

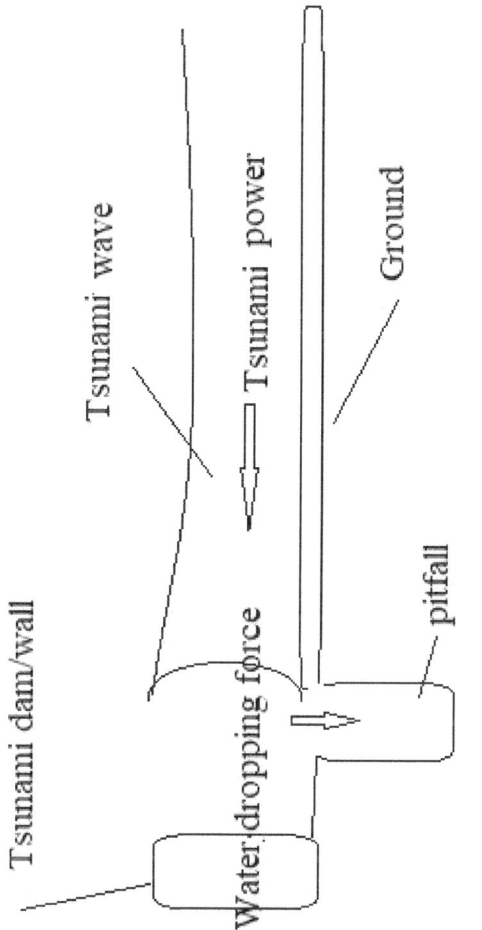

Stability through the Absorbed Water Even if the wave overcomes the wall, the wall remains intact and serves as a permanent obstacle. The absorption of water increases the weight of the wall, ensuring that it remains upright even during "flood" situations and doesn't topple over. The water absorbed from the wave would act as a dynamic stabilizing weight, helping to support the wall against wave energy. This added mass would prevent the wall from being pulled away or toppled by the wave's pressure forces.

Destabilizing the Wave The mechanism of water extraction from the wave could reduce its energy content, as less water mass would "enhance" the original wave. The wave buildup would be disturbed by the water withdrawal, potentially flattening the wave, as the energy is no longer transported in its original form. The wall ensures that the wave loses its original shape and intensity and that the water is drawn into the wall.

Advantages:

1. **Double Safety Levels:** Moat + Implosive Wall Opening: If the wave can overcome the moat, the wall is further stabilized by the implosion system and continues to absorb water. This provides a secondary line of defense.

2. **Stabilization through Water:** The absorbed water stabilizes the wall, even if the wave "overflows" or partially floods the wall. The added weight ensures that the wall does not tip over or get pulled away.

3. **Wave Energy Mitigation:** The wall has the ability to draw water from the wave, reducing the wave's energy and leading to a slowdown or flattening of the wave.

4. **Reduced Coastal Damage:** Even if the wave can surpass the wall, the intact structure of the wall will mitigate flooding and destruction on

the other side, as the water flow is reduced.

Possible Challenges:

1. **Efficiency of the Moat:** The moat needs to be deep and large enough to dampen enough water from the wave before it reaches the wall. The moat must work effectively to ensure that the wall is not flooded by the wave.

2. **Size and Durability of the Wall:** The wall must be very sturdy and capable of bearing both the high pressure from the wave and the additional weight of the water. If the wall is not properly dimensioned, it could be under too much pressure from very large waves.

3. **Risk of Blockage in the Water Absorption System:** It must be ensured that the water absorption mechanisms do not get blocked. Mud or debris could clog the openings and impair the system's function.

4. **Costs and Feasibility:** Designing such a structure requires high engineering skills and could be very costly, especially if implemented over large coastal sections.

Conclusion:

The two-stage moat-wave breaker wall with an implosive water absorption system presents an interesting way to reduce tsunami damage while maintaining a permanent barrier. The idea that the wall remains stable due to added weight and stabilization, even in the event of flooding, is clever, as it ensures the longevity and resilience of the structure.

Of course, this requires extremely complex technology and precise planning, but it could represent a promising solution to mitigate tsunami damage, especially when combined with existing protective measures.

Space for notes

other books:

	Titel	Medium	ISBN	Published
	Garden buildings made of wood - Terrace, seating area and similar	*Buch* *E-Book*	9783759731463 9783759710055	*10.06.24*
	Wooden garden construction - Animal Housing	*Buch* *E-Book*	9783757853525 9783757880200	*04.08.23*
	Broken Code	*Buch* *E-Book*	9783743192249 9783757836184	*03.04.23*
	Home Terra Preta - home made black soil	*Buch* *E-Book*	9783756231966 9783756245437	*29.06.22*
	Dan's Adventure in Africa	*Buch* *E-Book*	9783837059175 9783755794363	*25.02.22*

https://autorraginmund.de/buecher-und-romane-books

Publisher: BoD · Books on Demand GmbH,
In de Tarpen 42, 22848 Norderstedt, bod@bod.de
Print: Libri Plureos GmbH, Friedensallee 273,
22763 Hamburg
ISBN: 978-3-7693-9865-6